JOHN GAGG

CANALS

IN A

NUTSHELL

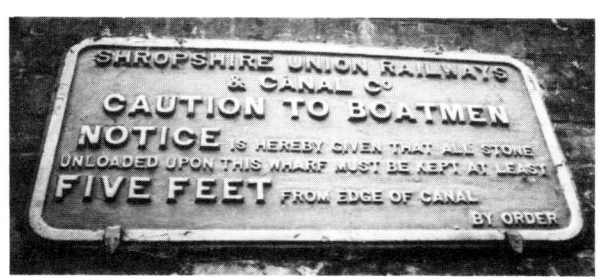

First published 1977
© John Gagg 1977
ISBN 0 9504226 4 9

Published by John Gagg, Shootacre House, Princes Risborough, Buckinghamshire
Printed by Manson Graphic, 12 Frogmore Road, Apsley, Hemel Hempstead, Herts.

Contents

This book is for a variety of readers. Many people, nowadays, are interested in canals, from those who walk their towpaths to those who cruise them in boats; from schools which study them to youngsters who paddle canoes. They see all sorts of intriguing things and often ask a variety of questions.

I have heard many of these questions, and it seemed to me that a simple introductory book about canals would be welcome. Most general waterway books are either large or technical or both, and they don't seem to tackle the subject through the eyes of someone actually looking at (and enjoying) canals on foot, by boat, or by car. So in this small book I have tried to sketch our canal system from this point of view, describing briefly the many aspects visible all over that system to those with eyes to see, and giving some of the basic statistics.

Obviously I can deal only in concise form with the fascination of canals, and more details about locks, tunnels, etc., appear in other books in this series. But I hope that this book is a useful potted account, and that it will be of value both to enthusiast and beginner.

The photographs are all my own, taken while cruising the whole waterway network over many years. The drawings are by Robert Wilson.

John Gagg

What is a "Canal"?

This isn't such an odd question as it sounds, for in fact when you come across a waterway it isn't always obvious what kind it is. Some of our waterways are pure canals, some are rivers, and some are varied mixtures of both.

Strictly speaking a canal is a waterway dug by man, usually for boats but sometimes for drainage. There are many hundreds of miles of these artificial channels, especially in the Midlands where there were few rivers suitable for boats. But even where rivers were in use to carry goods they were often shallow or swift in some places, so canal lengths were dug along parts of them.

Some of the rivers were almost turned into canals. The Aire & Calder Navigation, the Soar Navigation, the Calder & Hebble Navigation, parts of the Sheffield and South Yorkshire Navigation and a few others have long lengths of canal in them, with the river channel elsewhere. The word "Navigation" is especially used for this type of waterway, rather than either "Canal" or "River", but "Navigation" is also used in the names of some pure canals.

Even rivers such as the Nene and the Lee — and indeed almost all rivers used by boats — have canal cuts in them. Usually these are short stretches to and from locks while the river falls over a weir (a sort of shallow waterfall) nearby. But sometimes, as on the Trent through Newark, the canal cut may be much longer.

Thus it isn't always easy to decide at a glance whether a waterway with boats on it is a canal or not. The presence of locks is not necessarily a clue, for almost all navigable rivers, except for a few such as those of the Norfolk Broads, have locks on them just as canals have. Some, such as the R. Nene, have a large number, which help in flood control as well as navigation.

Some true canals do have their locks very close together in places where they climb slopes, whereas rivers are never so steep, and thus have their locks spread out. But so do most canals in most of their length. One noticeable point, however, is that rivers are normally wider than canals, and may have quite large weirs for the flow to bypass the locks and fall to the lower level. Canals have only small weirs to adjust their levels, or even none at all. But again, on some rivers the weir stream wanders off far from the lock, and the weir is out of sight. Some rivers, indeed (such as the Stort), are narrow and canal-like.

Anyhow, canals, rivers and "navigations" are mostly linked up in a network all over England and into Wales, with a few separate canals in Scotland. In England and Wales boats pass through the network quite happily from one kind of waterway to another. The chief thing to remember about rivers is that they flow along, are often deep, and can be much more dangerous than canals, especially near to weirs. But all water can be dangerous, not least around locks.

Thus this book will not deal with special aspects of rivers but, as I've shown, it is impossible to talk about canals alone, since so few of them are unconnected with rivers in any way. Even if a canal has no river as part of it, it may well join a river at one end or both, for that is one of the reasons why canals were dug — to link towns with the already-navigable rivers, or with other canals already linked with rivers.

You'll read mostly about canals, then, but don't be surprised when rivers pop up now and again.

Digging the Canals

The Romans dug canals in Britain, but almost all those that we have now were made in the second half of the 18th century and the beginning of the 19th. Although they now seem entirely part of the countryside, spare a thought for their digging.

There were no machines, of course, to carve them out, so men used spades and picks and wheelbarrows, with perhaps horses to help. They even had to burrow — sometimes miles — through hills in the same way. Yet as we cruise or walk now it seems unbelievable that men made the channel, bridges and locks, the mighty embankments, cuttings and tunnels, almost with their bare hands. These rough canal diggers were called *navigators* because they were building waterways, and this is where we got our word *navvy*.

The canal bed — usually invisible under the muddy water — hides many things, especially those thrown in by thoughtless people. But under everything it is lined with *puddle* — a mixture of the right type of soil and water to make it impervious. This was often trampled for hours by men to get it the right consistency.

The earlier canals stayed on the level for long distances by winding around the hills. The southern Oxford Canal shows this clearly. Later ones were made straighter by digging through the hills and using the soil to fill in the valleys.

The whole canal-digging mania, which spread to most parts of the country at one time or another, is almost beyond our imagining now. But people were eager to transport goods (before the days of railways, of course) more easily than on the appalling roads. So the navvies descended on remote villages and farms, often alarming people, then passed on their way after months of chaos. Yet all this is long forgotten, with the canals now some of the most peaceful places we have.

Canals Broad and Narrow

Canals, we often forget, were dug for boats to carry cargoes. The early engineers must turn in their graves when they hear Britain's canals compared with others in Europe now. For the sad fact is that our canals were dug for boats that proved too small in the end. Europe's canals carry huge barges, economically. Ours — even our biggest — will take only much smaller boats. We ought, of course, to enlarge them as Europe does, but that's another story.

The size of boat is determined by one main factor — the size of the locks on the canal. The depth of water is also important, but no boat can move any further if it's bigger than the first lock that it meets.

There have been all sorts of locks, from small ones for tub-boats up to ship-locks out to estuaries. But the many different canal companies realised that if they were to link up they would have to standardise their locks, in some areas at least. So in the midlands they agreed to a width of about 7ft, and a length of 70ft or a little more. This was really very small, but it saved on water lost down the canal every time a lock is used. These canals are now called *narrow canals,* and of course no boat can exceed the lock-size, however wide the actual canal might be. Boats 70ft by 7ft carry a very small amount of cargo compared with the barges of Europe, where 1350-ton capacity is common, and even bigger barges are in use.

Here is a boat in a narrow lock:

Elsewhere in England another lock-size, around 14ft by 70ft, became common, especially on the canals which now form the Grand Union, and the Chester and Erewash Canals, with a slightly lesser width on the Kennet & Avon Canal and the R. Wye. Canals with 14ft wide locks (or more) came to be called *broad canals,*

5

and two narrow boats from the narrow canals could conveniently fit side by side in a broad canal lock. This was called *breasting-up,* and they would sometimes travel long distances like this, being then the same width as the barges used on such waterways. When two narrow boats work together like this, one has an engine but not the other, which is called a *butty.* The motor boat usually tows the butty.

Broad lock, Grand Union.

Even broader lock, Aire & Calder.

Up north, a sort of breakaway group of canals made their locks 14ft wide or more, but shorter than the southern and midland 70ft. The Calder & Hebble locks varied, but officially take boats only 57ft 6in long. The Ripon Canal and R. Ure locks are only 57ft, but elsewhere in the north the locks are around 62ft long. One curious result of this is that the common 70ft by 7ft boats from elsewhere can't travel on these northern canals, although the locks are twice as wide as the boats are.

Some of the Yorkshire broad canals have quite large locks, taking big barges from Hull, for example. Trains of "Tom Pudding" containers, carrying coal to Goole, and the larger compartment boats feeding power stations, use long locks on the Aire & Calder Navigation and some locks leading to it from Doncaster. Long Sandall, an improved lock on the Sheffield & S. Yorkshire Navigation, is 215ft long.

The Manchester Ship Canal is exceptional in this country. Its locks are 600ft long, and the entrance lock is 80ft wide, to take ocean-going vessels.

Climbing up Hills

Locks have already been mentioned quite often, and without doubt these are the most interesting, varied and mysterious things on canals. Without them, except in absolutely flat country, canals couldn't exist. There are almost 1400 locks still in use on canals and rivers.

There's an entire small book about locks in this series, but this present book must include a potted section about these intriguing structures.

Why locks?

Water won't stay in a canal unless it's perfectly level. It would simply run down any slope and vanish. Yet apart from those in fenland areas, every planned canal came to a slope sooner or later. Some were made to climb quite high among hills and moors. Since the canal itself couldn't slope, it had to be made at different levels in turn, and boats had to be lifted or lowered from one level to the next. This, in fact, is the job that locks do. Men tried boat-lifts and slopes, and some of these are still in use abroad (with one lift left in England). But everywhere there are locks.

Men learnt about locks from the rivers which were used before canals. Rivers, of course, do slope, and water runs along them. But luckily it keeps on entering at the highest end. Dams had been put across many rivers to build up water to work mills or catch fish, or to make levels deeper for boats. The water above a dam deepened until it could pour over the top as a *weir,* or flow around through a mill-stream. Thus many rivers came to have different levels of water, with "steps" between them. This was why people first looked for ways of moving boats up and down these steps. Here is a weir by a former-mill on the R. Avon:

The earliest method of getting boats past weirs was to make some kind of opening in the dam that was holding up the water. Boards or a gate would be moved and the boat taken quickly through, then the opening closed again. But this lowered the depth of water in the higher level, and it was also quite difficult to drag a boat upstream against the strong current through the gap. Yet some of these water-gates, or *flash-locks,* could be seen until quite recently on the Thames and Avon.

Pound locks

The locks you see now are, in a way, two of these water-gates close together, with a "boat-box" in between. Their original name was *pound locks.* On rivers they are built near a weir. The water falls over the weir to a lower level but the boats go through the lock, instead of dangerously and wastefully through a single gap in the weir.

Because the lock is a sort of box with doors at each end (still called gates), the boat can change levels without a lot of water running away from the higher level. This also avoids the danger of going through a gap in a torrent of water.

Instead, water is let into (or out of) the lock until it is at the same level as the water outside in which the boat is floating. The gates at that end are opened, the boat goes in, and the gates are closed again. Water is then let out (or in) at the other end until the boat is again at the level of the water outside the other end. It has moved up or down as if in a lift. The other gates can then be opened and the boat moves out at the new level. The only water lost from the higher level has been one lock full, which makes hardly any difference.

You can see from this diagram how a boat is lowered from a higher level to a lower level of canal:

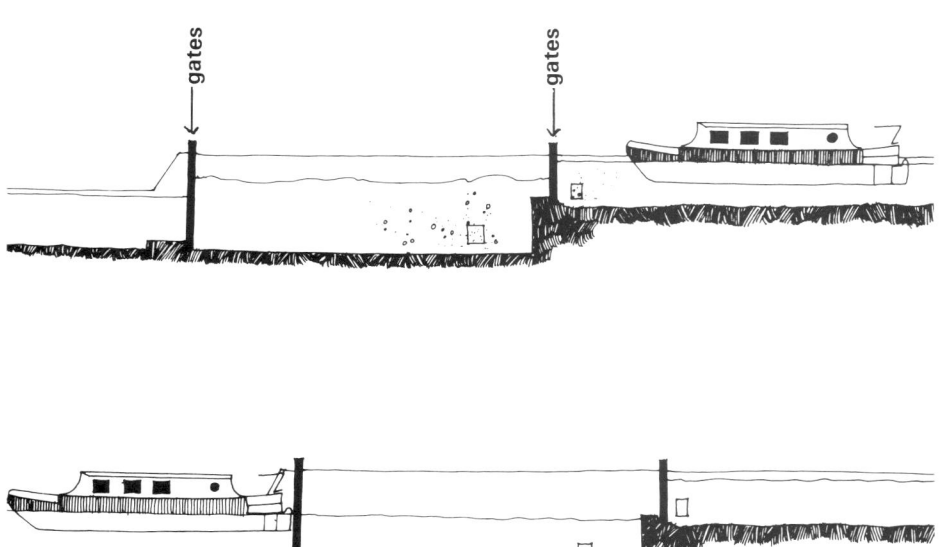

This sensible idea from rivers was a blessing for canal builders. They could take their canals over high hills by making many locks, provided they had enough water (from reservoirs at the highest level) to keep filling and lowering these locks.

To make sure that locks didn't use too much water each time, they were not made very deep, maybe raising or lowering boats about six feet. Thus on steep slopes there were many locks. Between Birmingham and Worcester there are 42 locks in five miles at one place, lowering the canal level 301ft. 23 locks near Wigan climb 214ft in two miles. In more level areas, though, locks are spread out.

Lock chambers

The "box" of a lock is usually oblong, and there are different sizes as already mentioned on ps. 5-6. A few locks are more like a diamond in shape, and there have even been round locks. You can see many of the normal oblong shapes in this book.

The sides may be of brick, stone, concrete, or steel sheets, or even cut in the solid rock. Some early locks had wooden sides. There are still occasional locks to be found (the Kennet & Avon, the Wey) with sloping grass sides. At each end of a lock chamber, below the water, there are solid stops called *sills,* of concrete or wood for the gates to close against.

9

Lock gates

Most gates are built of great wooden beams and planking, though there are also steel ones, especially in recent years. Almost all gates swing sideways, usually when you push with your behind against a *balance beam* which sticks out over the bank. These beams may be huge wooden ones or narrow metal girders. Sometimes there may be a handle instead, to turn to open the gate (at Blackburn, for example), or even a rope or chain to pull on. On busy commercial canals the gates are opened electrically by a lock-keeper in a cabin above the side of the lock.

On narrow canals there is normally one gate to close the top end of the lock. At the bottom end, where the gates must be deeper because the canal is lower beyond, there are usually two gates meeting in the middle. They meet at an angle against the water pressure of the lock. But you will find locks (on the Macclesfield Canal, for example) with two gates at each end; and elsewhere (Birmingham and the Oxford Canal, for example) with only one gate at each end. Broad canals have two gates at each end.

Very rarely on canals, but on several rivers, there are steel gates instead, lifting upwards with many turns of a handle.

Staircase locks

Where several locks are close together, with only short lengths of canal between them, they are called a *flight.* Sometimes, however, a slope is so steep that locks have to run straight into each other. They are then called *risers,* or a *staircase* of locks, and the gates between them are very tall.

Staircases can cause problems to boaters. Water from upper locks can flood the locks below, and of course boats cannot pass in a staircase of narrow locks. In broad locks it is possible for narrow boats to pass.

There are many staircases of two locks, and several of three. There are two groups of five at Foxton, with a short passing place between. There are five broad locks at Bingley in Yorkshire, and a staircase of eight on the Caledonian Canal in Scotland.

Foxton.

Some flights of locks		Lock staircases	
Tardebigge (41)	30	Banavie (9)	8
Lapworth (38)	25	Bingley (21)	5
Wigan (21)	23	Foxton (17)	(5
Hatton (17)	21	(two groups)	(5
Wolverhampton (5)	21	Watford (17)	4
Marple (29)	16	Northgate (35)	3
Stourbridge (37)	16	Grindley Brook (22)	3
Audlem (35)	15	Newlay (21)	3
Perry Barr (5)	13	Forge (21)	3
Farmers Bridge (5)	13	Field (21)	3
Rothersthorpe (17)	13	Bingley (21)	3
Bosley (23)	12	Oddy (21)	2
Aston (5)	11	Dowley Gap (21)	2
Wilmcote (38)	11	Stourport (36)	(2
Rushall (5)	9	(two groups)	(2
Napton (28)	9	Bascote (17)	2
Walsall, Ryders Green,		Marsworth (17)	2
Delph, Wheelock, ea.	8	Etruria (8)	2
		Bunbury (35)	2
		Brades (5)	2

(The numbers in brackets identify the canal, as listed on p. 32)

Lock Machinery

Because of the water pressure on one side, lock-gates can't be swung open until the water on the other side comes to the same level. That is why gates at both ends can never be open at the same time (except at special flood-locks on some rivers). Getting the water in or out of a lock to make these levels is the most fascinating aspect of the whole business of going through a lock. The water transfer is done by machinery usually called *paddle gear.*

The idea is simple enough. It is to uncover holes for the water to run through, and be able to cover them up again. But the ways of doing this are legion.

There may be water-holes in the lock-gates, or water-channels through the ground round the ends of the gates. These are blocked by shutters of some kind. In order to let water into or out of a lock, these shutters must be lifted or slid away.

We can't dive down to them, so we have to connect up with them by rods, and be able to pull or slide those rods somehow. At almost all non-electric locks this is done by turning a handle. Sometimes this is fixed on the paddle gear, but mostly you carry your own to fit on a *spindle*, and you lose it at your peril. This portable handle is called a *windlass*, or *lock-key*, and there are two common socket sizes (for two different sizes of spindle). But in places (Sheffield, Trent & Mersey) you really need a smaller socket still.

What happens when you turn the windlass varies greatly, since different canal companies devised different paddle gear. The Leeds & Liverpool Canal has the greatest variety of all. It would be an enormous shame if all this fascinating gear were standardised, yet this is happening in places.

Various cog-wheels turn as you wind, often moving a *rack* full of teeth upwards, thus pulling up the shutter below. Some paddle gear works easily, some is very stiff. With some you stand precariously on a step on the gate. With others you may have to fix the windlass on a spindle right at the end of the balance beam, almost kneeling to do it. With some you face one way, with some the other. Some turn one way, some the other. Some ground paddles send a small water-spout up your leg.

The Leeds & Liverpool, among its variety, often has sideways-sliding boards covering its water-passages. As you wind you move a rack sideways instead of upwards, to swivel the board. At other Leeds & Liverpool locks you turn a handle horizontally, drawing up a paddle on a worm-drive. Simplest of all, you heave up a wooden arm to slide the board below water.

The Calder & Hebble has a remarkable system. Here you carry a great square-ended wooden *hand-spike,* and insert it in a socket again and again to lever up a rack.

The least intriguing system is the newest, working hydraulically. Everything is safely hidden from view, and you wind at great length as a tiny indicator rises up an enclosed box. Unlike the older paddles, it takes as much energy, and as long, to wind the paddle down again.

The old paddles, indeed, will drop with a crash if you're careless, and this is bad for them (but useful in an emergency). In any case, to keep them up you make use of a variety of ratchets, designed to lock the cogs until you are ready to release them and wind the paddle down again. Needless to say, these gadgets don't always work. And most of them are oily, and can easily trap fingers.

You can see a variety of paddle-gear opposite, from different canals.

13

Lock Furniture

Besides the essential gates and paddle gear, there are various other things associated with locks, and found at some or all of them.

Crossing locks

Somehow those who work the locks must be able to cross from one side to the other. There may be a road bridge, but more often there isn't. Special separate footbridges are often built at one end, especially on the Leeds & Liverpool and parts of the Trent & Mersey Canals. Sometimes such bridges have a gap in the middle for the former horse tow-ropes. Some footbridges on the Staffs & Worcs Canal are beautifully shaped. Occasionally a swing bridge crosses the lock and may have to be moved to allow boats to pass.

There are almost always footplanks fastened to the gates, sometimes wobbly, and often these are the only ways to cross. This means that the boat crew may have to walk to the closed end of the lock to cross. But on narrow canals agile lock-workers jump between footplanks when one of a pair of gates is open and the other closed.

Ladders and steps

Deep locks on broad canals may have ladders let into the walls so that you can climb into or out of a boat. Some lock gates have rungs or footholes for precarious climbing. Outside the lock there may be worn old steps down the slope to the lower canal level, or in a few cases even a ladder.

Bollards

Many locks provide *bollards* near the locksides for tying boat-ropes. Some are ancient gnarled wooden ones. The Grand Union and other canals offer very solid metal ones. The Calder & Hebble has curious curved ones. More modern bollards everywhere are of concrete. There are bollards in the wall of Gloucester lock, and in some Lee Navigation locks.

It is dangerous to tie up a boat tightly if it is going down in a lock, but using lock bollards can prevent boats from crashing about as water is let into a lock.

Bollards are also provided above and below some locks, to tie up boats while the lock is being prepared.

15

Kebs

These are the long-handled rakes occasionally seen, and used for extracting all sorts of junk from the canals. They will fish out lumps of timber, oil-drums, bricks and other debris that jams behind lock gates, as well as pulling out reeds and piles of mixed rubbish that may have floated down the canal and ended at the lock. Many a keb has fished out people and animals, too.

Lock cottages

Most lock cottages have vanished, and lock-keepers are now few. But some houses remain, either with lock-keepers or privately owned. On the southern Stratford Canal there are some beautifully kept distinctive ones, and as with many things, each canal has its own pattern.

Mysterious clumps of roses and lilac at many locks betray a once-cultivated garden of a vanished lock house.

Walkways

Generations of boaters have heaved on balance beams, walking the same tracks and pushing with their feet. Often this has made a shallow trench, full of muddy water, but some canals provide a curved brick or concrete walkway for the beam-pusher. The ideal ones are ribbed, with perhaps a special rib at the other end to give the first strong purchase.

Stop-planks

At some locks (and also at some bridges) you will see a set of stop-planks, often under a shelter, or even in a hole under a bridge. These can be dropped in turn down slots in the lock walls outside the gates, to fit across the canal. They then form dams to hold back the water. The water between them is pumped out and men can get at the gates or anywhere else in the lock to do repairs. Stop-planks are also used to seal off longer lengths of canal when necessary.

Working a Lock

The business of getting a boat through a lock is usually called "working the lock". The only real way to learn how to do it is to do it, but it is possible to do it for years and still get in a muddle over it. Most waterway guides and hire-firm booklets give details, and a good hire firm should give a demonstration. In this small book I can't go into every move, but I have already mentioned several tricks of the trade and shall now mention a few more. A pictorial guide to working a narrow lock is given in the companion **Book of Locks.**

The chief jobs are concerned with the proper use of paddle-gear and a proper respect for the power of water. Paddle-gear can be damaged, and it can damage you. Water can crash a boat about, damage gates, and of course drown people. So you use gear carefully, and never start letting water into or out of a lock until you're quite sure that the boat is under control and the gates are closed.

If the water in the lock isn't at the same level as that of a boat wanting to enter, it must be brought to that level by emptying or filling. That's what the paddle-gear is for. But the boat must be kept away from the strong water-movement either way, and you must make sure that the paddles at the other end are fully closed.

Having got the levels right the gates can be opened and the boat taken in. Then the paddles must all be closed, and also the gates.

The opposite process then happens. With the boat under control (but not tied tightly if it is "going down"), the water-level is changed by opening the opposite paddles. There are two dangers if the lock is being filled. The boat can be drawn uncontrollably forward; and also, if there are paddles in the gate they must not be opened too soon, or water may gush in and sink the boat.

If possible, keep the boat away from the gates anyway. For even when a boat is quietly dropping down in an emptying lock, its propeller could sit down on the sill of the gate behind it.

When the lock is at the new level, gates are opened, paddles closed, and the boat moved out. Gates are closed again, and hey presto! — the job's done. But oh! what a performance some people make of it!

Broad locks are more trouble with narrow-width boats. The swirl of a filling lock can tear a boat across with a bang. Usually it is helpful to open the ground paddle on the same side as the boat, and the water bouncing in holds the boat against the wall. But you can't be sure, and it's especially necessary to secure boats in such locks.

There's often no need to open both gates at the end of a broad lock. Ordinary narrow canal boats can go through one gate, halving the labour. .

There are many hints and tips to add to the bare details. Some were mentioned in the descriptions of paddle-gear, bollards, ladders, staircases, footbridges, and so on. You make sure no-one is on the wrong side of a swinging balance beam, or standing on the rope which is holding a boat. You don't leave a windlass loose on a spindle, in case something slips. You have a bicycle with you if there are lots of locks. And you don't "steal" a lock by emptying or filling it for yourself when someone is coming the other way who could use it first. (Anyhow, they then do the work for you!)

One thing is certain. Even after years, there are still odd tips to be picked up on lock-working. But some of those used by the old boatmen are too cruel to the shakier gates and gears of today.

Canal Boats

There are various boats in photographs in this book because boats, of course, are the reasons why canals exist. The original boats were commercial ones, apart from the maintenance boats and the occasional luxury inspection boats of the canals' directors. Nowadays commercial traffic is seen almost only on some of the broader-locked canals and rivers, with a few enthusiasts at work on narrow canals.

In Yorkshire coal is carried to power stations and for export, with other cargoes also on the same canals. There is trade around London, and some in the north west, with coasters on the Gloucester-Sharpness and Manchester Ship Canals. Besides barges, the Yorkshire canals have the container-trains mentioned on p.6.

Elsewhere, apart from the occasional trading narrow boat (often selling coal along the canal) there are large numbers of pleasure boats. These range from canoes to 70ft long pleasure versions of the commercial boats, built on the same lines but with cabins instead of cargoes.

Some are in fact conversions of trading boats, in all their slightly-differing varieties. Others are specially built. They have living quarters as comfortable as those in a house, with beds, cooker, fridge, hot and cold running water, shower or even bath, flushing lavatory, coal stove or gas heater, and electric light.

Shorter boats may still have the traditional "narrow boat" shape, or they may be of "glass reinforced plastic" (GRP), and look more like boats seen on rivers or the sea. They may still have many comforts, even if only 25 ft long or less. The smaller boats need fenders, or strong strips along their sides and bottom, to protect them through locks and narrow bridges, or against junk under water.

Many people own such boats, keeping them at special mooring places in marinas, at boatyards, or even just at the canal side where this is allowed. The boats must have licences for the canals, and fees are paid for keeping them at the moorings.

Large numbers of pleasure boats are hired out — usually for a week or a fortnight — by people who want to cruise on canals and rivers. There are hire firms all over the waterways. You steer and run the boats yourself (the firm should show you how). If you would rather be lazy, there are even hotel boats on canals, often taking up to twelve people in small cabins on two boats, one towing the other.

Besides commercial and pleasure boats, there are also the occasional *maintenance boats,* working on the canal. These may be small flat boats with men cutting hedges or towpaths, or bigger ones with tools and machinery for repairing banks, bridges or locks.

Canal banks need to be "piled" sometimes — metal sheets knocked into the ground to stop the bank from washing away. Special *dredgers* scoop out mud where the canal has become too shallow, and may put it into *hoppers* to be taken away by *tugs.*

Elsewhere in this book there is much about boats moving along canals — in locks and tunnels, across aqueducts and under bridges. Steering a boat — and especially stopping it — and mooring it and looking after the engine are not hard to learn. But they do need a bit of common sense.

Bridges

Everywhere along canals there are bridges, for every road and farm track that a canal meets must cross over it (or go under it).

Different canal companies built their bridges in different styles, and of course with local material. Midland canals mostly have warm brick hump-back bridges, often now worn in places. There are massive stone bridges in the north and in Wales, with those on the Macclesfield a distinctive shape. They curve inwards a little at the bottom, helping to keep the cabin roofs of boats from catching their slopes; whereas some canals (the Oxford, the Shropshire Union) have bridges with awkward slopes at one side to scrape boat-roofs. The Leeds & Liverpool and the Lancaster have solid stone bridges, and surprisingly the Ashby has too.

Normally, of course, the towpath curves and goes under the bridge too, so that the canal doesn't pass exactly under the middle. But some companies economised by omitting the towpath. This must have annoyed boatmen, who would have to unhitch the horse. But on the Stratford Canal a gap was left in the middle of the bridge for the tow-rope, rather like those in the lock footbridges on the western Trent & Mersey.

The eastern Trent & Mersey, however, has many brick bridges at the foot of locks with no towpath, and even one near Burton far from a lock. Some of these bridges (Middlewich, Alrewas, for example) are very worn where boats have scraped their way in and out of the locks.

Towpaths change sides along canals, and often a special bridge was built for this job. It enabled the horse to walk up a slope, cross over, and walk down another slope towards the boat, then turn under the bridge and continue on the new side. These are called *turnover, roving,* or *snake* bridges. There are also towpath bridges at junctions, including some fine metal ones on the Birmingham Canals.

Tow-ropes chafed bridge corners, so metal posts were often put there. Even these usually now have deep grooves in them from generations of ropes (see p. 30).

One fascinating aspect of canal bridges is their numbering. Though some numbers have disappeared, most bridges have a number-plate — beautiful ones, with names also, on the Staffs & Worcs Canal, simpler ones elsewhere. Leeds and Liverpool numbers often have letters after them — 103, 103A, 103B, etc., — and some bridges elsewhere have carved numbers. With the aid of the canal guides, this is often the boater's only clue to his whereabouts.

Although few bridges show their names, almost all possess one — no doubt with a story behind it. Milking Hill, Uncle Ben's, Ugly, Clogger, Giggetty, Cuckoo, Adam & Eve, Paradise and Long Moll's are among the intriguing ones to be found in the guides.

Thus there is variety everywhere — in the materials, shapes, widths and even heights, for on the Shropshire Union and near Birmingham, for example, high bridges span cuttings. But bridges in other places may hardly leave room for boats.

There are also movable bridges. Usually these are small country or farm bridges, though the Leeds & Liverpool has a few on busy roads. On the commercial canals in Yorkshire, and along the Gloucester & Sharpness Ship Canal, bigger bridges are lifted or swung by bridge-keepers, sometimes electrically.

Simple lift-bridges are found on the Oxford, Llangollen, Caldon, Brecon, Peak Forest and northern Stratford Canals. Some have high beams, Dutch-style. Swinging bridges appear less frequently, except for over 50 on the strenuous Leeds & Liverpool. There are odd ones on the Coventry, Grand Union, Macclesfield and Peak Forest.

Sometimes movable bridges are stiff with dirt, age, or abuse.

Tunnels

When the canal diggers had climbed fairly high to cross hilly country, they had a problem. Should they continue to build more locks to get over the summit, and if so, how would they get water into the highest level?

The answer usually resulted in a tunnel through the highest part. In this way the number of locks (which delayed boats) was reduced, and reservoirs could be made at higher levels to feed the canal. The engineers also made tunnels through smaller areas of high land on the routes of canals.

Cowley (Shropshire Union)

With their picks and shovels and wheelbarrows, and with dangerous gunpowder, the navvies burrowed deep below ground, often in terrible conditions. For long tunnels, several shafts were sunk from the hill-top, after the line of the tunnel had been marked by stakes. The workers were lowered down these shafts, and dug outwards from them. Soil and rock was taken up the shafts in containers on ropes, maybe pulled by horses.

Blisworth tunnel, 3056yd long, had 19 shafts. Shafts were often lined with brick and left for ventilation later. You can see "chimneys" on the Blisworth shafts now, and even the mounds where the soil was left.

Men were killed at this work, and often the navvies struck rock, or loose sand, or underground springs. Many tunnels took much longer to dig, and cost much more than the canal companies expected. Some are rather crooked, too, with being dug from so many separate places at once. There are many stories of the different tunnels, and more details of the 46 still in use (six over a mile long) and of the main closed ones, are given in **Canal Tunnels**, a companion book in this series. This also describes the remarkable task of "legging" boats through a tunnel — for few had towpaths.

When cruising through a long tunnel now, the boater often enters the tunnel mouth after a long cutting. Usually, once inside, it is just possible to pass a boat coming the other way, but a few tunnels are too narrow. You need a headlamp, and maybe even an umbrella, for many tunnels drip water from their roofs.

The walls may be of varying colours because of water, chemicals, and even smoke from long-gone steam engines. The light of an oncoming boat creeps slowly up, until at last you squeeze carefully past, unable to see anything of each other. The engines thunder from the walls, and the distant end is a pinhole at first, or even invisible on dull days. The whole journey is an eerie experience, and the far exit is like another world.

Braunston (Grand Union).

Tunnels still in use

Canal	Location	Length (yd)	Canal	Location	Length (yd)
Ashby	Snarestone	259	**Leeds &**	Foulridge	1640
Birmingham	Dudley	3154	**Liverpool**	Gannow	559
Canal	Netherton	3027	**Llangollen**	Chirk	459
Navigations	Gosty Hill	557		Whitehouses	191
	Coseley	360		Ellesmere	87
	Galton	122	**Monmouthshire**		
	Ashted	113	**& Brecon**	Ashford	375
	Summit	103	**Oxford**	Newbold	250
	Curdworth	57	**Peak Forest**	Hyde Bank	308
Caldon	Leek	130		Woodley	167
	Froghall	76	**Rochdale**	Deansgate	78
Chesterfield	Drakeholes	154	**Shropshire Union**	Cowley	81
Grand Union	Blisworth	3056	**Staffs & Worcs**	Cookley	65
(main)	Braunston	2042		Dunsley	25
	Shrewley	433	**Stratford**	Kings Norton	352
(Leicester)	Crick	1528	**Trent & Mersey**	Harecastle	2926
	H. Bosworth	1166		Preston Brook	1239
	Saddington	880		Barnton	572
(Regents)	Islington	960		Saltersford	424
	Maida Hill	272	**Worcester &**	West Hill	2726
Kennet & Avon	Bruce	502	**Birmingham**	Shortwood	613
	Bath No 1	59		Tardebigge	580
	Bath No 2	55		Dunhampstead	230
				Edgbaston	105

Aqueducts

Besides climbing up hills — or going through them — canals also came to streams and rivers and had to cross them. This they did on water-bridges — aqueducts.

There are many hundreds of aqueducts, but most of them are over little streams which take quite a small tunnel under the canal. From the canal itself you may not even notice them. But at other places the waterway may cross bigger rivers or wide valleys, and occasionally roads.

The engineers hoped to keep the canal on the level to do this, rather than taking it down by locks and up again, as this would waste water. So in some places aqueducts are quite striking as they stride across from one side of a valley to the other, perhaps on an embankment for some way before the "water-bridge" starts.

The famous Pontcysyllte aqueduct on the Llangollen Canal has an embankment 1500ft long at one side of the valley, 97ft above the original ground level just before the aqueduct starts. The aqueduct itself carries the canal in an iron trough of 418 plates, 1007ft long, on 18 tall pillars. At one point the River Dee is 121ft below. The waterway is 11ft 10in wide, but there is a towpath over part of this, so the normal 7ft wide boat only just fits. A crossing of this aqueduct can be an alarming experience in a wind.

A solid acueduct near Burton on the Trent & Mersey Canal is more squat, and often not noticed by boaters since the canal is quite wide. Yet it has nine brick arches over the R. Dove, and 14 other arches on the approach — 1¼ miles altogether.

Among other striking river-crossings, a massive stone aqueduct takes the Lancaster Canal over the R. Lune, the Dundas aqueduct takes the Kennet & Avon over the Avon, and an unusual semi-suspension one carries a branch of the Aire and Calder over the Calder at Stanley Ferry.

Great Ouse (Grand Union)

Edstone (Stratford)

The most astonishing aqueduct of all is one at Barton which takes the Bridgewater Canal over the Manchester Ship Canal. Its predecessor was also a great attraction, for when Brindley built it of brick over the Irwell, people didn't believe it possible. They would certainly be staggered by the modern steel one, which swings open, full of water, to let ships pass below. The ends of a tank — and of the canal — are sealed off by swinging gates, and the whole 1250-ton weight pivots to the side of the Ship Canal channel.

Aqueducts over roads may require only a small tunnel for the road, though there are also notable examples of this type. A tall, impressive stone aqueduct takes the Macclesfield over a road at Bollington, and one takes the Grand Union over the North Circular in London. Motorists between Birmingham and Stratford are surprised to see boats above them at Wootten Wawen, as are others passing under the Shropshire Union at Nantwich, and also on the A5 at Stretton.

Perhaps the most intriguing aqueducts are where narrow canals cross narrow canals. The Macclesfield crosses the Trent & Mersey in this way, after leaving it earlier, and the Leek branch of the Caldon crosses the main line. There are three such "flyovers" near Birmingham.

Some of the main aqueducts still in use

Canal		Canal	
Aire & Calder	Stanley Ferry (Calder)	Llangollen	Ponycysyllte (Dee)
Ashby	Shenton		Chirk (Ceiriog)
Birmingham	Engine	Macclesfield	Bollington
Canal	Steward		Dane
Navigations	Spouthouse		Red Bull
	Ryland	Monmouthshire &	Brynich (Usk)
	Piercy	Brecon	Gilwern (Clydach)
Bridgewater	Barton Swing	New Junction	Don
	Mersey		Went
	Bollin	Peak Forest	Marple (Goyt)
Caldon	Hazelhurst	Shropshire Union	Stretton (A5)
Chesterfield	Retford (Idle)		Nantwich (A51)
Coventry	Tame		Weaver (2)
Grand Union	Wolverton (Ouse)	Staffs & Worcs	Sow
	North Circular		Trent
	Rugby (Avon)		Stour (2)
Kennet & Avon	Dundas (Avon)	Stratford	Edstone
	Avoncliff (Avon)		Yarningale
Lancaster	Keer		Wootten Wawen (A34)
	Lune		Cole
	Wyre	Trent & Mersey	Croxton (Dane)
	Brock		Rugeley (Trent)
Leeds &	Priestholme (Aire)		Dove
Liverpool	Dowley Gap (Aire)		
	Douglas		

Canalside Pubs

Waterway inns are a vast field of study, even if only from the outside. Obviously their original purpose was to sustain the commercial boatmen, and perhaps their horses, on their journeys. With the decline of trade many pubs must have closed, unless they could turn to roads for customers. But now we are seeing a little of the opposite drift, with pubs that have ignored their canals for years once again providing moorings and canal decoration, and even in some cases changing their names.

Pub names are usually watery. *Navigations* win the day, with *Bridges* close behind, including special ones such as *Napton Bridge, Bridge House,* etc. There's many a *Wharf,* though not as many *Locks* as you'd expect, but *Top Lock, Big Lock, Kings Lock* and others can be found.

There are *Swans* and *Boats* (and *Ferry Boat, Longboat, Narrow Boat, Pleasure Boat*), and a surprising number of *Anchors* considering the lack of such items on canals. Surprisingly, too, there are *Ships* around, but not as many *Barges* as you'd think.

Johnson's Hillock locks, Leeds & Liverpool Canal.

There are a few special names to do with aqueducts, junctions, and specific canals, though what a *Jolly Tar* is doing in Cheshire is a mystery. Then of course there are plenty of non-canal names, with the usual *Lions* of different colours, and well-known canal pubs called the *Plum Pudding,* the *Blue Bell,* the *Ash Tree* and the *Malt Shovel.*

Some of these pubs along canals are tiny and little changed. Some are smothered with juke-boxes and plastic now. If you are a pub addict you need to stick your nose carefully in the door first, for you may prefer to retreat.

Keep your eyes open!

Walking along a towpath, or cruising at the same speed in a boat, means that there's a lot of time to observe things along a canal. Compared with driving a car all kinds of items impress themselves on the mind, simply because it takes so long to pass them. Even so, those who never notice anything anyway don't notice canal things either. So keeping your eyes open is a rewarding habit.

Obviously much of what there is to see is at locks, tunnels, aqueducts, bridges and pubs, and many eye-catching items have already been mentioned. But here are a few more to be spotted.

Notices and numbers

Canals produce all sorts of notices — especially the Shropshire Union. Modern ones tell you how to work locks, and ancient cast-iron ones tell you not to moor over stop-grooves (Ellesmere), not to tip stone (Llangollen), and not to drop lift-bridges. The Shroppie also bars traction engines "or other extraordinary weights" at one bridge, and the Kennet & Avon has some fascinating notices at Newbury lock, including one about horses not crossing the street. There are water-tap notices, and many bridge-notices for road vehicles, which no-one obeys. The Waterways Museum has some fine preserved notices, including threats of transportation which might well be revived for present-day vandals.

Bridge-numbers and their usefulness were mentioned earlier, and on some canals there are lock-numbers, too. Solid metal ones can be found on Stratford and Llangollen Canal gates in places, and the Kennet & Avon boasts some, with an excellent wooden imitation at Lock 81.

Some of these numbers merely count the locks in a flight, and there are remarkable examples carved in all the entrance walls of the 23 Wigan locks (in Roman numerals), and in places on the Wolverhampton flight of 21. The Tardebigge locks also have carved numbers in the walls, and many smaller flights have numbers on their balance beams.

The Birmingham Canals have an interesting system of numbers on the buildings scattered around the network. A lock-house on the Birmingham & Fazeley Canal, for example, has the number 258, with the date MDCCXX.

Mileposts

Many delightful mileposts remain. Large numbers of unusually-shaped Shropshire Union ones have been rescued from the undergrowth, and quite a few distinctive Trent & Mersey posts still mark the distances between Preston Brook and Shardlow. A few old Grand Junction Canal Co. plates stand on the Grand Union, and the Brecon has mileposts with just numbers on them. The Leeds & Liverpool, typically, has many massive milestones, while there are odd ones on the northern Coventry and the Peak Forest, with some fine oval plates on the Lancaster. Try also the Chesterfield, the Gloucester & Sharpness, and the distance-plates in Blisworth tunnel.

Other posts

All along the Ashby Canal there are markers in the fields. They say "MR" on the opposite side, showing the later canal-owners — the Midland Railway — staking its boundary claim.

Such boundary posts are not uncommon. The Brecon has some elaborate ones, the Coventry has many, the Staffs & Worcs has the occasional one (try Tixall lock), and I saw one on the Middlewich branch of the Shropshire Union. Presumably all canal companies once marked their territory opposite the towpath in some way.

A different kind of post is the distance-post (above, centre), occasionally found on the Oxford and Grand Union Canals, for example. This marked a given distance each side of a lock, and the boat which came to its post first had priority in that lock. On the River Weaver tall concrete posts, marked "200" (yards), served the same purpose. The other post above is a fine example of a bridge corner protection against towing-ropes.

Canal buildings

Lock houses, mentioned earlier, were the most common buildings connected with canals, and some still remain. But there were various other canal buildings. Bridge-keepers on the commercial canals have their houses, with pillared ones on the Gloucester & Sharpness especially attractive. Old stables can often be found, as at Bunbury, and at the mouths of some tunnels.

Toll houses have sadly faded from the Birmingham Canals, and it is difficult to track down examples anywhere else now. But warehouses still stand in a number of places — Sheffield, Sowerby Bridge, Audlem, for example — in varying states of revived use or decay. Present-day boatyards have rescued many such buildings, as well as old wharves and basins.

Maintenance yards and buildings still operate, of course, some well kept like Wigan, Hartshill, Bulbourne, Ellesmere and Northwich.

Birds, flowers and animals

Canals offer an especial wealth of natural science. No one should cruise or walk them without at least a flower book and a bird book.

Water-birds are the most common, obviously, and most canals boast at least one aggressive swan, and other gentler ones. Moorhens are everywhere, with bundles of chicks in spring. Ducks (and drakes), dabchicks and coots are there, with herons slowly taking off and landing ahead of you, to stand gazing into the water. Kingfishers are harder to spot, but in winter especially their blue flash is common, skimming ahead. Magpies are around always (raise your hat to single ones), and wagtails too, with many non-water birds preferring the quiet of canals.

Every imaginable flower can be found, for the dampness attracts the ordinary ones as well as the water ones. Everything grows more lusciously, from rosebay willow-herb to meadow sweet. Water-lilies appear on some canals (even at Slough), and yellow irises — especially on the Ashby. Reedmace (wrongly called bulrush) is frequent, and the flowers of spring — primroses, violets, anemones, bluebells — line some banks.

Animals are fewer. Water-voles (not "rats") plop and swim, but for the rest you must be thankful for a sight of rabbits or hares in the fields, or a rarer dusk glimpse of a badger or a fox. Of course there are always sheep and cows in fields alongside.

Canals and Navigations still in Use

		Miles	Locks
1	Aire & Calder Navigation	41½	17
2	Ashby-de-la-Zouch Canal	21¾	0
3	Ashton Canal	6¼	18
4	Beverley Beck	¾	1
5	Birmingham Canal Navigations	106¼	139
6	Bridgewater Canal	39¾	1
7	Calder & Hebble Navigation	21½	37
8	Caldon Canal	20¼	17
9	Caledonian Canal	60	27
10	Chesterfield Canal	26	16
11	Coventry Canal	38	13
12	Crinan Canal	9	13
13	Erewash Canal	11¾	15
14	Exeter Ship Canal	5	1
15	Fossdyke Canal	11¼	1
16	Gloucester & Sharpness Ship Canal	16	1
17	Grand Union Canal (all sections)	250	275
18	Huddersfield Broad Canal	3¼	9
19	* Kennet & Avon Navigation	86½	104
20	Lancaster Canal	45¼	6
21	Leeds & Liverpool Canal	141½	104
22	Llangollen Canal	46	21
23	Macclesfield Canal	27½	13
24	Manchester Ship Canal	36	5
25	Middle Level Navigation (and Old Bedford River)	91¾	7
26	Monmouthshire & Brecon Canal	33¼	6
27	Nottingham Canal (and Beeston Cut)	5	3
28	Oxford Canal	77	44
29	Peak Forest Canal	14½	16
30	* Pocklington Canal	9½	9
31	Ripon Canal	1¼	1
32	Rochdale Canal	1¼	9
33	Selby Canal (with R. Aire)	11¾	4
34	Sheffield & S. Yorks Navigation (and New Junction)	48½	29
35	Shropshire Union Canal (and Middlewich Branch)	76½	50
36	Staffordshire & Worcestershire Canal	46¼	45
37	Stourbridge Canal	5¼	20
38	Stratford-upon-Avon Canal	24½	55
39	Trent & Mersey Canal	93½	76
40	Witham Navigable Drains (approx)	87	2
41	Worcester & Birmingham Canal	30	58

* parts of these canals are not yet restored.
(Locks to tidal waters are not included unless in normal inland use)